ESSENTIAL SCIENCE

D0462321

the expanding universe

MARK GARLICK

SERIES EDITOR JOHN GRIBBIN

LONDON, NEW YORK, MUNICH,
MELBOURNE, AND DELHI

senior editor Peter Frances
senior art editor Vanessa Hamilton
US editor Cheryl Ehrlich
DTP designer Rajen Shah
picture researcher Sarah Duncan
illustrator Richard Tibbitts

category publisher Jonathan Metcalf
managing art editor Phil Ormerod

Produced for Dorling Kindersley Limited by
Design Revolution Limited, Queens Park Villa,
30 West Drive, Brighton, East Sussex BN2 2GE
editors John Watson, Sarah Doughty
designer Katie Benge

First American Edition, 2002

02 03 04 05 10 9 8 7 6 5 4 3 2 1

Published in the United States by
DK Publishing, Inc.
95 Madison Avenue
New York, NY 10016

A Cataloging in Publication record is available from
the Library of Congress.

ISBN 0-7894-8416-1

Color reproduction by Mullis Morgan, UK
Printed in Italy by Graphicom

see our complete product line at
www.dk.com

contents

introducing the universe

How big is the Universe? Let's try to put it into human terms. The fastest jet fighter planes can cover more than 0.6 miles (1 kilometer) per second, three times faster than sound. Even at this speed, it would take one million years to reach the nearest star beyond the Sun, Proxima Centauri. And yet, if this distance were reduced to the size of a flake of oatmeal in your breakfast bowl, the furthest galaxies would be on the other side of the planet Earth. On such scales, it might seem mind-bogglingly unrealistic that astronomers claim to know so much about the cosmos and all of its contents. Still, modern researchers have many tools to help them in their quest to unravel the mysteries of space. The last century saw more developments in science and technology than had occurred in the whole of history. In this book, you will find out where the Universe came from and where it might be going. But first, we look at what exactly is in the Universe – and find out how astronomers know what they claim to know about it.

death star
This image, taken by the Hubble Space Telescope, shows the death throes of a star as it casts off a shroud of gas, known as a planetary nebula.

a short tour of everywhere

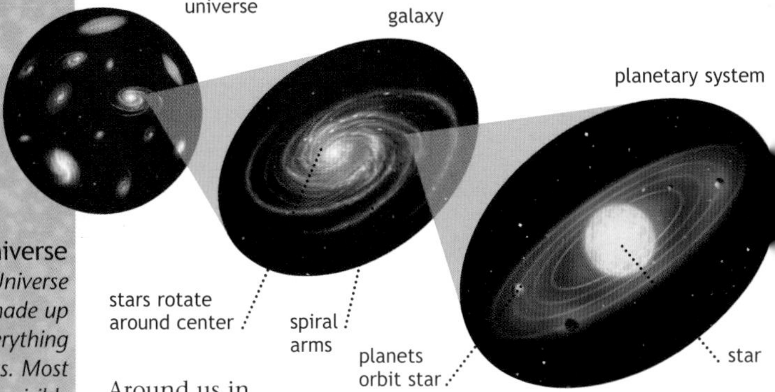

universe

galaxy

planetary system

stars rotate
around center

spiral
arms

planets
orbit star

star

the universe
*The Universe
is made up
of everything
that exists. Most
of the visible
matter in the
Universe is
grouped into
countless galaxies.
These are made
up of stars and
planets forming
planetary systems.*

Around us in
every direction is a Universe of planets, comets, stars,
galaxies, nebulae, and gas and dust clouds. On a dark,
clear night you will probably see several thousand stars,
a planet or two, and a few cloudy areas. One of those
cloudy areas is another galaxy – Andromeda – a gigantic
island of stars and the furthest and largest object you can
see without optical assistance. Andromeda is 2.9 million
light years away and is more than 100,000 light years
across. In the context of the Universe, Andromeda is
nearby. At the other end of the scale, astronomers
measure distances in billions of light years. We'll now set
out on a tour to see what else is out there, starting with
the objects closest to home: the planets.

planets

Before 1800, the only planets that were known were the
innermost six of the nine planets that make up the Solar
System. However, astronomers now know that planets are

common, and probably exist throughout the Universe.
Planets come in two varieties. The smaller ones are
called terrestrial planets. They are made mostly of rocky
and metallic substances, and
have rigid surfaces, but they
may or may not have an
atmosphere. Mercury, Venus,
Earth, Mars, and possibly Pluto
all fall into this category. The
other planets – Jupiter, Saturn,
Uranus, Neptune, and all the
planets so far found around
other stars – are several times larger and are known as gas
giants, even though they are not really made of gas. They
are made up of substances that on Earth naturally exist as
the gases hydrogen and helium, but within the giant
planets themselves they actually exist in liquid form. So
gas giants are really vast globes of spinning fluid. All of
them have atmospheres that blend into their interiors,
and they probably have solid cores.

> **"Place three grains of sand
> inside a vast cathedral, and
> the cathedral will be more
> closely packed with sand than
> space is with stars. "**
>
> Sir James Jeans, British
> mathematician (1877–1946)

stars

The vast majority of planets orbit stars, just as Earth
revolves around the Sun. Even with the most powerful
telescopes, most stars appear no larger than dots. In
reality, stars are massive, hot gaseous globes that are tens
or even hundreds of thousands of miles across. They
occur in all sizes and colors – even combined in pairs,
which orbit one another and are known as binary stars.

**large
magellanic cloud**
*At 179,000 light
years away, the
Large Magellanic
Cloud is one of the
nearest galaxies to
us, and contains
more than a billion
stars. The blue
cloud to the left
was created by the
explosion of a
massive star.*

star birth and death

Stars are born in vast clouds of gas and dust called nebulae, and the nebula shown here contains stars of all ages. The dark clouds in the top right-hand corner have not yet started to collapse to form stars. The pillars of glowing hydrogen at the bottom right contain emerging stars. The bright cluster contains young massive stars whose lives will be short; while the blue supergiant at top left is close to star-death.

At the lowest end of the stellar scale are the smallest and most common stars, known as red dwarfs. Red dwarfs are typically half the mass of the Sun and have surface temperatures of around 7,000°F (4,000°C). Sunlike stars are slightly hotter, yellow, more massive, and not so common. At the highest end of the scale are the very luminous stars, over ten times more massive than the Sun. These are blue giants. They are extremely rare and exceptionally hot, with a temperature of more than 90,000°F (50,000°C). However, all of these stars burn in the same way throughout their lives. But when stars get older, they undergo some astonishing changes. For example, the Sun will eventually begin to die by first swelling up to become a stellar monster known as a red giant, which is many hundreds of times larger than a conventional star although much lighter. This stage is followed by collapse and death, leaving behind a tiny corpse known as a white dwarf, which is a hundred times smaller than a normal star.

nebulae

The clouds of gas and dust called nebulae, in which
numerous stars form, are mainly huge reservoirs of
hydrogen and helium, but they also harbor traces of other
gases and some grains of carbon with icy surface coatings.
Nebulae are either bright or dark depending on how they
are viewed and whether they are near any stars. The light
from nearby stars reflects off the gas to produce a
reflection nebula, or it makes the gases in the nebula glow
like an aurora and we see a bright, so-called emission
nebula. But if no stars are nearby, the gas does not reflect
light and instead remains dark, visible only by any bright
matter that it partially obscures. The largest nebulae are
the giant molecular clouds. They can be hundreds of light
years across and contain enough material to make
millions of stars.

galaxies

Larger still, though, are galaxies – vast
islands of nebulae, stars, and, it is
presumed, planets. Galaxies occur in
three basic forms. The Milky Way is an

spiral galaxy

example of a spiral galaxy and contains
approximately 200 billion stars. As the
name suggests, spiral galaxies have spiral-
shaped distributions of stars and nebulae,
and are generally flat like a disc. However,
they also have a central bulge so that from
the side they look like a pair of fried eggs

elliptical galaxy

back to back. The largest galaxies are the
elliptical galaxies. They can be many times
more massive than spiral galaxies and well
over 100,000 light years from one side to
the other. Elliptical galaxies are shaped like
giant footballs. But unlike footballs,
the three axes have different lengths.

irregular galaxy

types of galaxy
*Galaxies have
been classified
according to their
shape. The two
main types are
spiral galaxies
and elliptical
galaxies. Galaxies
that are neither
spiral nor elliptical
are described
as irregular.*

Elliptical galaxies are also different from spiral galaxies in that they contain very little nebular material, and therefore very few new stars. Lastly, there are the so-called irregular galaxies, which conform to neither of the other two classes. However, not all irregular galaxies are as misshapen as their name suggests. Some of them are roughly disc-shaped and, like spiral galaxies, contain active regions of star formation, but they have no discernible spiral arms.

galaxy clusters

Just as stars gather together under the influence of gravity to form galaxies, so gravity in turn makes galaxies group together into much larger clusters. The largest clusters, such as the Virgo Cluster, contain thousands of individual members and occupy regions of space about 20 million light years across. But small clusters, like the so-called Local Group that includes both the Milky Way and Andromeda, house only around 30 or so relatively small galaxies in a region that spreads about 5 million light years across. Generally, like galaxies themselves, the richest galaxy clusters show some kind of structure, with very massive – usually elliptical – galaxies dominating the clusters' central regions. Here in the central regions, clusters can be extremely dense, with galaxies only a few diameters apart from each other – much, much more closely packed than stars. But further

distant galaxies
These widely differing galaxies, found in the Hubble Deep Field image, are the furthest ever seen, and are forming relatively shortly after the beginning of the Universe.

out from the cluster core, the density drops off and the galaxies themselves become smaller, irregular, and more widely spaced, containing only a few million stars each.

superclusters

Galaxy clusters are still not the largest structures known to exist. For just as galaxies tend to congregate together, so do entire clusters – gathering to form truly vast assemblages known as superclusters. On the very largest of scales, the Universe has a foamlike consistency to it. Enormous sheets of clusters and superclusters – each harboring billions of galaxies – form the walls or filaments of the "bubbles" in the foam. Inside the bubbles are huge empty spaces called, appropriately enough, "voids" – which themselves can be 150–200 million light years across. Almost all of the visible matter in the entire observable Universe is locked up inside these gigantic sheets and filaments. Despite the billions of galaxies that exist, much of the Universe looks incredibly empty. Indeed, there seems to be only one feature larger than superclusters – and that is the Universe itself. The Universe is as big compared to the largest asteroid as that asteroid is compared to the tiniest subatomic particles, called quarks.

billions of galaxies
A computer-generated slice through the Universe shows the organization of billions of galaxies in clusters, which are linked by slender filaments crossing billions of light years. The darker areas are vast regions of empty space.

dark regions are empty space or voids

bright areas are walls or filaments composed of superclusters of galaxies

how we know what we know

How do astronomers know what they know? How can they tell how far away the stars are, how big they are, how much mass they contain, and so on? The answer has a lot to do with the equipment that researchers use. But important clues also come from the way in which many astronomical objects behave and interact.

photometry

One of the most basic activities anyone can do in astronomy is watch an object to see how its light levels change over time. This science is known as photometry – literally, "measuring light." For example, think of an asteroid rotating in space. Asteroids are irregularly shaped lumps of metal or rock, smaller than a planet. An asteroid shaped like an egg will look brighter when seen from the side when more of it is in view, than when seen end-on. So, merely by studying how the light from the asteroid changes over time, astronomers can tell how fast it rotates and form some idea of its shape. Now imagine a star that shows very slight changes in brightness over a period. This could be a sign that planets are orbiting the star because the light of the star would dip very slightly as the planets pass in front of the star and slightly obscure it. Two stars may be orbiting one another, or alternatively, a star might have spots on its surface. As the star spins, its brightness

light effects
In these two illustrations, the effect on light output by two binary stars orbiting one another is shown. No two stars are identically luminous, so the light levels change during each orbit.

darker star in front reduces the light

brighter star increases the light level

history of the telescope

Galileo (1564–1642) made the telescope famous by constructing a refracting telescope containing two lenses that refracted, or bent, the light that entered it. In the 1670s, English physicist Isaac Newton (1642–1727) developed a better design of telescope based on mirrors to make the first, working, reflecting telescope. Both types of telescope continued to be used, and by the early 20th century, larger telescopes were possible. Another use of telescopes is to "hear"

Isaac Newton's telescope

the universe. Many objects in space give off strong radio signals, and radio telescopes began to be built to detect them. However, visible light and radio waves are only a part of the electromagnetic spectrum, and more information can be gathered by examining the radiation that comes from other parts of the spectrum usually obscured by the atmosphere. In the 1940s, a telescope was proposed to be positioned above the atmosphere. Eventually the Hubble Space Telescope was launched in 1990. All its instruments are installed behind the main mirror, and a smaller mirror reflects images back into this area of the telescope for recording and analysis.

eagle nebula

hubble pictures
The Hubble telescope became fully operational in 1993 and took this image of the Eagle Nebula in 1995.

Hubble Space Telescope

would vary depending on how many darker, spotty regions are in view at any given moment. These very small photometric changes can now be detected and suggest that planets, starspots, or other stars are present.

the electromagnetic spectrum

Electromagnetic radiation is a form of energy that travels through space and matter. Different types of radiation, distinguished by their wavelength and energy, collectively form the electromagnetic spectrum. Only a narrow band of radiation in the center of the spectrum is visible to the human eye, and this we call light. So, to "see" more of the universe, telescopes have been built to detect objects that emit very little light radiation, but emit other electromagnetic wavelengths. The first alternative telescopes were radio telescopes, and then others were developed to detect other electromagnetic wavelengths.

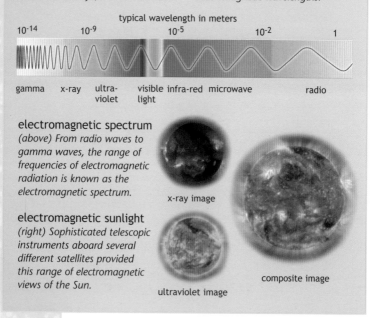

typical wavelength in meters

10^{-14} 10^{-9} 10^{-5} 10^{-2} 1

gamma   x-ray   ultra-violet   visible light   infra-red   microwave   radio

electromagnetic spectrum
(above) From radio waves to gamma waves, the range of frequencies of electromagnetic radiation is known as the electromagnetic spectrum.

electromagnetic sunlight
(right) Sophisticated telescopic instruments aboard several different satellites provided this range of electromagnetic views of the Sun.

x-ray image

composite image

ultraviolet image

spectroscopy

As useful as it is, photometry nevertheless has its limits. A more powerful technique is spectroscopy. Light is passed through a series of narrow slits and is split up into a spectrum. The spectrum is broken up by dark "spectral lines." These lines occur because the light source is made up of atoms that absorb light at specific wavelengths associated with particular colors. A particular element absorbs light at its own particular range of wavelengths. For example, spectral lines at one range of wavelengths can only mean that helium is present in the star, while spectral lines at a different range point to a different material. This method allows astronomers to conclude which gases are contained in the objects they are looking at. Moreover, the exact range and intensity of each atom's spectral lines also varies with their physical properties. Spectroscopy not only indicates what the objects are made of, but also how hot or dense they are.

using spectroscopy
The colored spectrum is broken up by dark spectral lines where the light has been absorbed by atoms of specific elements. Pressure, density, and temperature can be measured by the widths and intensities of the individual lines.

.....spectral line

the doppler effect

Another feature of spectroscopy is that it can reveal how fast objects are moving. Imagine a fire engine, sirens wailing, coming toward you. Incoming sound waves from the fire engine are compressed because of the forward motion. This gives the sound a shorter wavelength, which creates a higher pitch. When the fire engine has passed, those same waves are now stretched out while on their way to you so they have a longer wavelength, which

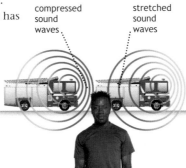

compressed sound waves

stretched sound waves

light effects
By applying the Doppler effect to the light emitted by stars and galaxies, it is possible to figure out whether they are moving toward or away from us.

creates a lower pitch. This is the Doppler effect. The precise frequency of the sound that you hear is directly dependent on the speed and direction of motion of the fire engine in relation to where you are. This is important in astronomy because the same phenomenon works with light waves. A star moving toward you has its light waves compressed so that its spectral lines will appear at a higher frequency – a little bluer than if the object was stationary relative to you. This is known as blueshift. Similarly, a redshift is seen if the object moves away from you. The wavelengths of spectral lines can tell astronomers which way the object is moving and how quickly.

judging distance
It is easy to judge the relative distance of street lights. Far-off lights appear dimmer than close ones. In space it is hard to tell if a bright star is a faint one close by or a luminous star far off.

distances

If astronomers know how bright a star is intrinsically – that is, how much light it actually gives out – they can estimate how far away it is. By analogy, if you know how bright a car's lights are up close, you can tell how far away the car is from how bright its lights seem.

Astronomers rely on a class of variable stars whose average intrinsic luminosities are known well, the so-called Cepheids. These stars have light levels that go up and down at precise periods lasting a few hours. The key is that the brighter any given Cepheid is intrinsically, the slower it varies in brightness. So the period of variability, easily measurable with photometry, instantly tells astronomers how bright the star is up close. And just as with the car's headlights, if you know how bright the star is in reality, you can figure out – from its apparent brightness – how far away it is.

how to weigh stars

To weigh a pair of binary stars, astronomers need to find out their speed and how long it takes the stars to complete an orbit (the "orbital period"). The law of gravity directly relates this period to the masses of the two objects and to their distance from each other. All objects in the Universe attract one another. The attraction between the Earth and an amusement park car can be measured by weighing the car, which is the pull of gravity between the car and the Earth. This depends on, first, the mass of the car and the mass of the planet. If the weight of people in the car is doubled, gravity pulls on it twice as hard. Second, the further the car is from the center of the

falling not floating
The mass of this amusement park car and its distance above the Earth define the car's weight and its speed under gravity. Similarly, the distance between two orbiting stars and their speeds can reveal what their masses are.

planet, the weaker the pull exerted by the planet; this force weakens by a rule known as the inverse-square law. Using this law with the orbital velocity of the stars as revealed by spectroscopy, astronomers can estimate the distance between two stars. And from that, they can calculate their masses – in a sense, they can weigh stars.

the inverse-square law
Gravity follows the inverse-square law: when the distance between two objects is doubled, they attract each other with one-quarter of the strength. If you triple the distance, the force drops to one-ninth. So the effect of gravity decreases by the square of the distance.

object 1

object 2

distance between the objects

distance between the objects is doubled

gravity decreases by a factor of four

how it
all began

Although the Universe is far larger than the mind can ever comprehend, it is principally made up of countless stars passing through their cycle of being born, living, and dying. These stars gather into vast groups called galaxies, which are grouped into mighty galaxy clusters. Immense though it is, the Universe was once squeezed into a tiny speck with essentially no physical size, but with an infinite pressure and density. Suddenly there was an unimaginable event called the Big Bang, after which the Universe began to take shape as a vast cloud of hydrogen and helium gas, expanding at huge velocities and cooling from temperatures impossibly hotter than that of the Sun. Some astronomers believe that in less than one billion years, the first stars and galaxies formed from these materials. As these first stars came to the end of their lives, and exploded and died, the first atomic elements heavier than lithium – and from which planets and people are made – were blown into space. The details of how this happened are still a subject of much debate.

the old big bang
This detail of an image of the Universe shows the "leftovers" of the Big Bang. It is the background radiation left behind and is possibly 13 billion years old. This radiation is partly responsible for the "snow" that you can see on the screen of an untuned television set.

the big bang

the beginning
The Big Bang was the fireball of creation that created the basic materials of the Universe – and was the start of time itself.

It is the biggest question anyone could ask in astronomy: where did the Universe come from? The answer, it seems, is from nowhere. An overwhelming amount of evidence shows us that the Universe is expanding, growing larger by the second. So, running the clock backward suggests that long ago the entire Universe occupied a single point, known as a singularity, which then somehow expanded enormously to produce the Universe we see today. If this Big Bang theory of the Universe's sudden emergence is correct, and

a dissenting voice...

It was the British scientist Sir Fred Hoyle (1915–2001) who coined the term Big Bang in the 1940s to describe this theory of the birth of the Universe. Hoyle had actually used the expression as a sarcastic rebuke – he was strongly opposed to the idea right up to his death. Nevertheless, much to his surprise and somewhat ironically, the term stuck fast. No other theory for the origin of the Universe has had anything resembling this success. The observed expansion of the Universe, the presence of cosmic background radiation, and even its distribution across space, all strongly support the idea that the Universe sprang from a single point between 12 billion and 15 billion years ago.

if astronomers have accurately measured the rate of expansion of the Universe, then it all began about 12–15 billion years ago.

fireball of creation

The seed from which the Universe sprang was unimaginably small, dense, and hot. All the matter that we see around us today was squeezed into a volume of space far smaller than a subatomic particle. Perhaps this conjures up an image of a tiny, hot, and dense speck floating in a cosmic void, waiting to explode. But there was no void. The fundamental features of the Universe – space, time, and

The American astronomer **Edwin Hubble** (1889–1953) was the first to prove that galaxies – until his time believed to be types of nebulae – are actually vast islands of stars. However, his most astonishing discovery was that the observed galaxies were moving away from each other – and that the Universe is therefore expanding. The Hubble Space Telescope is named after him.

matter – were tied up inside the singularity, and so there could be no "outside." Neither can we say that the cosmic seed exploded to produce the Big Bang – again, there existed no dimensional space for it to explode into. Similarly, the notion of "before" the Big Bang defies the boundaries of logic because the Big Bang created time itself. All we can say is that first there was a cosmic seed, which for reasons we may never understand, suddenly began to expand – into a fireball of creation.

inflation

In the very first instants, the expansion of the fireball proceeded very slowly, but then began to speed up. If this rate of expansion had not continued to increase, the Universe would have collapsed back in on itself within a fraction of a second, and space and time would have vanished with it. But the Universe did not collapse because it underwent a startling change in the rate of its expansion – a period that cosmologists call inflation. As the minute fireball expanded, it also cooled rapidly in the same way that gases cool when their volume is increased. At about 10^{-35} seconds after the Big Bang (that's a decimal point followed 34 zeros and a 1), when the Universe was still far smaller than a subatomic particle, the temperature reached a crucial 10^{28} degrees. This resulted in a tremendous amount of energy being released, which blew the young Universe up by a factor of 10^{20}–10^{30} in around 10^{-32}

> **"In the beginning the Universe was created. This has made a lot of people very angry and been widely regarded as a bad move."**
>
> Douglas Adams (1952–2001), British writer, *The Hitchhiker's Guide to the Galaxy*

antimatter

There is nothing bizarre about antimatter except that it is rare. Every species of subatomic particle has its counterpart antiparticle, which is its mirror opposite. The negatively charged electron has an associated antiparticle with an equal but opposite electric charge – called a positron. Even electrically neutral particles have antimatter counterparts, in which other fundamental properties are reversed. Antimatter and matter particles existed in nearly equal quantities in the early Universe.

But because they destroyed each other on contact, most were annihilated almost immediately. The slight surplus of matter that remained now makes up the Universe around us.

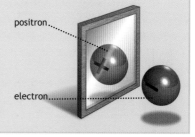

positron

electron

seconds. In this brief event, which lasted about a trillionth of a second, the subatomic Universe grew to the size of a healthy melon. This equates to increasing the size of one atom to millions of times the diameter of the Solar System in the time it takes a ray of light to cross one-billionth of the width of a single subatomic particle.

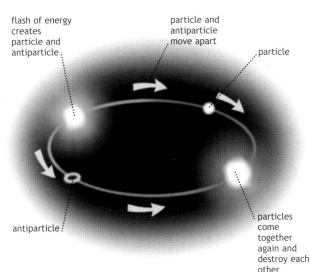

flash of energy creates particle and antiparticle

particle and antiparticle move apart

particle

antiparticle

particles come together again and destroy each other

brief lives
In the earliest stages of the Universe, particle and antiparticle creation occurred all the time. In a burst of energy, a particle would be created along with an antiparticle. The two would exist for a brief instant of time before coming together again and annihilating one another in a further burst of energy.

matter from energy

After the cessation of inflation, the Universe continued to expand, but at a considerably subdued rate. The Universe at this time was a seething, boiling cauldron in which energy transformed into matter and, in turn, matter was transformed into energy. Some of the radiation energy (photons) changed into matter in the form of subatomic pairs of particles and antiparticles. As soon as they appeared, these particle-antiparticle "virtual" pairs came together and annihilated each other to change back into radiation. However, not all of the particles were annihilated. Owing to slight imbalances in the very early

Universe, there were very slightly more particles than antiparticles – there were about a billion and one particles for every billion antiparticles – and this had huge consequences later on.

the first stable particles

As the Universe grew larger, so it became still cooler and its photons less energetic. The particles that briefly came into existence had less mass and therefore less energy. Finally, there came a point where the Universe lacked sufficient energy to make any new particles. The particle-antiparticle pairs that existed at this point changed back into energy for the last time and the Universe became a much quieter place. But not all the matter vanished – those one-in-a-billion particles that had no antimatter equivalents with which to mutually destruct were left stranded. These became the first stable subatomic particles that still exist today as the building blocks that make up atoms: the protons, neutrons, and the electrons. If the early imbalance in the cosmic fireball that favored particles rather than antiparticles hadn't existed, these building blocks would not exist and the Universe would not contain its galaxies, stars, planets, and people.

hydrogen nucleus

proton

helium nucleus

two neutrons

two protons

atomic nuclei
All atoms consist of a nucleus made up of neutrons and protons. The exception is the light element, hydrogen, which has only a single proton as its nucleus. The nucleus of an atom of helium is made up of two protons and two neutrons.

the first atomic nuclei

With the Universe almost three minutes old, the remaining particles in the Universe made up an incredibly hot gas. In this gas, the protons, neutrons, and other particles constantly collided and bounced off one another because the temperature provided the particles with the energy to travel at immense speeds. As the clock ticked past three minutes, the temperature cooled to about over 1.8 billion °F (1 billion °C), and some of the protons and neutrons now collided relatively gently and were able to stick together to form tightly bonded clusters that make up the hearts of atoms – atomic nuclei.

Most simple atomic nuclei are combinations of protons and neutrons. The nucleus of a helium atom, for example, is constructed of two protons and two neutrons. The hydrogen nucleus, however, consists of a single proton and no neutrons.

> **❝ Water, air, fire, and earth, these are simply different clusters of the changeless atoms. ❞**
>
> Democritus, 439 BC

These are the most basic elements; the third simplest element is lithium. These three – hydrogen, helium, and lithium – were the first nuclei to be created. Few other elements made an appearance. While the Universe was hot, it contained enough energy for the background radiation photons to produce numerous particles that could stick together to make helium and lithium. But when the temperature dropped, fewer particles were produced, and so few elements heavier than lithium could be created. Additionally, the expansion of space also ensured that collisions were becoming less frequent.

the first stable atoms

Although the period of particle-antiparticle annihilation was over and the Universe contained stable matter, it was

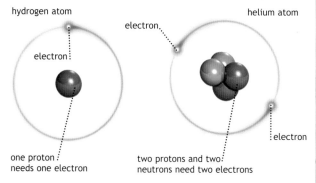

hydrogen atom

electron

electron

one proton
needs one electron

helium atom

electron

two protons and two
neutrons need two electrons

electron

atomic structure
An atom is formed only when a nucleus – made up of a proton, or a combination of protons and neutrons – is orbited by a number of electrons equal to the number of protons. The creation of stable atoms proved very difficult in the early Universe.

only in the form of a bath of "ionized" gas of atomic nuclei, electrons, and photons. There were no atoms because nuclei need electrons in orbit around them to form whole atoms. But every time an electron connected with a nucleus to make an atom, a highly energetic photon would collide with the electron. The photon's energy would be absorbed by the electron, which would then have too much energy to remain in orbit around the nucleus and would fly off. This process is known as ionization, and it continued unabated for the first several hundred thousand years of the Universe's existence.

In the 1960s, using a radio antenna at Bell Laboratories in New Jersey, two physicists named **Arno Penzias** (b.1933) and **Robert Wilson** (b.1936) found, quite by accident, that the sky was full of microwave radio "noise." They came to realize that this was the redshifted remnant of the Big Bang itself – its "echo." In 1978 they were awarded the Nobel prize for physics for this invaluable contribution to cosmological knowledge.

Another feature of the Universe at this stage, when the first stable atoms were about to be formed, is that it would have been impossible to see through space. We can see because light in the form of a stream of photons stimulates the cells at the back of the eye. But early on, no photon could travel very far without having its energy absorbed by an electron, and, without the photons, nothing could be seen. However, after roughly 300,000 years the temperature cooled to about 11,000°F (6,000°C), and many photons then lacked sufficient energy to ionize. Electrons could join nuclei without being knocked off, and the first complete atoms of hydrogen and helium appeared. Photons were able to travel through space for the first time. and the Universe became transparent to light. The Big Bang was over.

key points

• First stable subatomic particles appear when the Universe is 10^{-4} seconds old.

• At three minutes old they organize themselves into the first atomic nuclei.

• After about 300,000 years, the first whole atoms form.

background radiation

The point at which the Universe became transparent to light is extremely important. The radiation that had filled the Universe with a sea of photons no longer interacted strongly with matter, so the radiation could roam the Universe forever. The Universe expanded, matter reorganized itself into galaxies and stars – and the background radiation remained as a feature of the new Universe. However, it did not remain unaltered because as the space it occupied expanded, so did the wavelength of the background radiation. The idea that space itself has been expanding since the Big Bang is one that can be difficult to understand. However, it is still happening and anything passing through space, such as the background radiation of the Big Bang, is also still being stretched. In fact, the wavelength of the background radiation now corresponds to the microwave (or short-wave radio) region of the electromagnetic spectrum.

blue = -454.7°F (-270.40°C)

pink = -454.81°F (-270.45°C)

Astronomers have detected this radiation by pointing radio antennae into space. They found that its "temperature" – a measurement of the energy of the radiation – is a few degrees warmer than absolute zero. Detecting this radiation has been hailed as one of the most important confirmations of the Big Bang Theory.

map of the sky
This map of the Universe reveals tiny differences in the temperature of the background radiation throughout space. The pink regions are about 0.01% warmer than the average temperature of space, and the blue regions are a similar fraction cooler.

longer waves
Imagine a line in the shape of a wave drawn on the surface of a rubber band. When the rubber band is stretched, the wave stretches as well. In the same way, space has expanded and so the background radiation that fills space has been stretched as well.

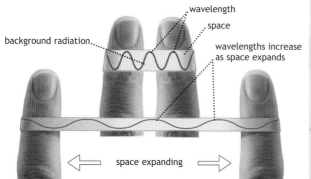

background radiation

wavelength

space

wavelengths increase as space expands

space expanding

edge of a galaxy
Images of the Universe as it is today reveal extraordinary objects, such as NGC 4013 – a spiral galaxy that just happens to be perfectly edge on to Earth.

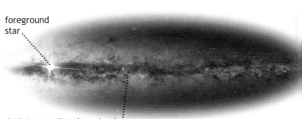

foreground star.

dark interstellar dust clouds

the universe today

Observations have shown that temperature differences indicate differences in the density of matter in certain regions of the Universe. Areas of slightly higher density had stronger gravity than low-density areas, and it was in these regions of slightly higher gravity where the first galaxies emerged between one million and one billion years following the Big Bang.

However, there must be more matter than we are aware of in the present Universe for galaxies to hold together and move through space in the way that they do. In particular, the way in which galaxies rotate cannot be fully explained simply on the basis of the detectable matter that they contain. The stars furthest from the center of galaxies rotate at a greater speed than could be generated simply by the mass of discernible matter. One of the theories that attempts to explain this is known as dark matter theory. This suggests that there is matter that outweighs the bright stuff by about ten-to-

missing matter
This swarm of stars is called a globular cluster. The way in which these clusters orbit our galaxy suggests that there is more matter in the Universe than we currently know of.

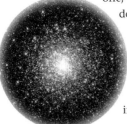

one, but cannot be seen because it does not emit any form of electromagnetic radiation. It is truly dark. Some think that dark matter exists in the form of hypothetical subatomic particles called WIMPS – weakly interacting massive particles.

birth of galaxies

Scattered throughout the cosmos like dinner plates a few meters apart, galaxies are vast collections of stars held together by gravity. Some are flattened discs, others oval-shaped, and still others have no discernible form at all. Even the smallest galaxies harbor around one million stars, and the largest are a million times larger. Altogether, it is very possible that there are more of these cosmic islands in the Universe than there are stars in the Milky Way. But the question of where these vast structures first came from is still unanswered. Did they form as dust clouds before stars formed, or did the stars appear first? Faced with questions like these, the best that astronomers can do currently is speculate.

the pre-galaxy era

Astronomers are not even certain when galaxies started to form, let alone how. Some researchers say that it was a million years after the Big Bang, while others prefer a

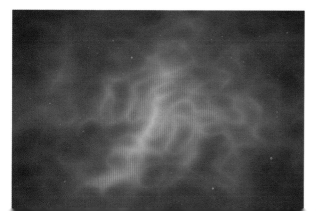

galactic sites
It has been discovered that in the early Universe these denser areas of matter with greater gravitational attraction formed the regions in which galaxies were to develop.

The English philosopher **Thomas Wright** (1711–86) of Durham was forward-thinking in his 18th century view that the Milky Way is a vast island of stars all moving in the same direction, with the Sun displaced from the center and the galaxies situated outside it. It wasn't until the 20th century that more modern scientists proved him to be broadly correct.

figure closer to a billion years. The best estimate is that galaxies began to form at some point between these two extremes. Happily, there are some things that are known with more certainty. Before galaxies existed, the Universe, fresh from the Big Bang, was a cosmic cloud of hydrogen and helium gases, mixed in with much larger quantities of dark matter.

Like real clouds, the early Universe was patchy, with some areas slightly denser than others – a direct result of the period of inflation. The denser regions, having a stronger gravitational attraction, gradually pulled in matter from their surroundings, and the local density grew greater still. After a while, the cloud began to develop into discrete collections or filaments of tightly packed matter, with smaller traces of material in between.

the milky way

There are two main ideas about galaxy formation, but most observational evidence seems to support a "bottom-up" scenario (see panel, right). The nearest evidence can be found right here in our galaxy, the Milky Way. The Milky Way is essentially shaped like a disc, but it is surrounded by a gigantic spherical "halo" of star clusters that swarm around it like wasps around their nest. The stars within these so-called globular clusters have ages that differ from each other by several billion years. Astronomers know this because as stars age, they become redder. Some of the stars are redder than others, so these

galaxy-formation theories

The events that led to the division of the early Universe into galaxies after extended filaments of matter had been created are uncertain. There are two theories. One suggests that the Universe formed from large, dense gas clouds that fragmented.

Each became a single galaxy through gas clouds contracting under gravity to form stars. According to this theory, galaxies grow from the top down, whittled away from much larger volumes of matter.

top-down theory

large region of primordial gas cloud in the early Universe

primordial gas cloud fragments, and each fragment begins to collapse

single galaxies form through gas clouds contracting under gravity to form stars

Another theory argues that galaxies grew from the bottom up. It supposes that the initial dense regions in the Universe were relatively small – far too small to

subdivide into structures as large as entire galaxies. These regions began to merge with their neighbors through gravity to make larger structures and, eventually, galaxies.

bottom-up theory

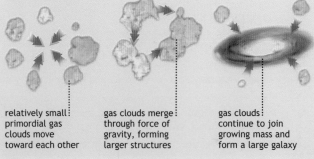

relatively small primordial gas clouds move toward each other

gas clouds merge through force of gravity, forming larger structures

gas clouds continue to join growing mass and form a large galaxy

individuals must be older. This means that the globular clusters, and therefore our whole galaxy, probably did not form at the same time, as would have happened in the top-down scenario. Instead, the Milky Way formed over several billion years as it gradually sucked in more and more gas clouds, which then condensed into stars.

galaxy cannibalism

Whether galaxies formed from the top down or from the bottom up, there is now clear evidence that they have evolved greatly since they came into existence, and that they have done this by interacting with each other. This is hardly surprising, because galaxies are much, much closer together in relation to their size than stars are. As a result, collisions between individual galaxies were, and are, common. The largest galaxies of all, the so-called giant ellipticals, are so large – containing up to ten times as many stars as the Milky Way – that they probably evolved by swallowing their smaller neighbors. Some of these galaxies have internal structures that suggest they are still "digesting" the galaxies they swallowed earlier. The best indication that galaxies have evolved

star story
The colors of stars indicate how old they are because young stars tend to be bluish, and older stars become yellow, orange, and red. This view of predominately red and orange stars in the Milky Way indicates that our galaxy probably formed billions of years ago.

Harlow Shapley
(1885–1972) was the first person to gauge the scale of the Milky Way accurately by measuring the distances to the spherical clusters of stars that orbit it – the globular clusters. He showed that the galaxy was much larger than had been thought. Later, he produced a catalog of galaxies that revealed how they were spread across space – grouped into vast clusters.

through cannibalism can be found in images from the Hubble Space Telescope, which produced the Hubble Deep Field image (see page 10). The galaxies it shows are billions of light years away, which means that we are seeing those galaxies as they appeared billions of years ago. Astronomers are able to compare these ancient galaxies with more recent galaxies to fill in the time gaps and figure out what might have happened in between. In the Deep Field image, spiral galaxies are much more common than elliptical ones; but in images of closer galaxies, ellipticals are more common. This is interpreted as meaning that there were more spiral galaxies billions of years ago, before they had time to collide with one another and merge to form today's elliptical galaxies.

galaxies collide
The gravitational forces of the larger galaxy on the left have distorted the shape of the smaller galaxy, which in billions of years' time will be completely absorbed by its larger neighbor.

galaxies today

Currently there are an estimated 50–100 billion galaxies that can be identified with modern telescopes, but the total number of galaxies is likely to be truly immense.

As far as existing galaxies are concerned, astronomers estimate that about 60 percent of them are elliptical, 30 percent of them are spiral galaxies, and around 10 percent of them are irregular. The cannibalism process is continuing, and the ratio of elliptical galaxies to spiral galaxies will increase in the future.

life of stars

Galaxies are made of stars, and because we live inside the Milky Way galaxy, stars are scattered across space in every direction that we look. Stars come in many different sizes, temperatures, and colors, but they all originate from the same source – cold clouds of gas and dust, that cover an area many light years across.

giant molecular clouds

The raw materials for making stars are spread across our galaxy and others in immense patches of interstellar fog known as giant molecular clouds, or starbirth nebulae. The Orion Nebula is a famous example. These clouds typically measure dozens of light years across. They are made primarily of hydrogen gas in which the individual atoms are joined together in twos to make simple molecules – molecular hydrogen. The remaining material, about one-quarter by mass, is mostly helium, but there are also traces of carbon, oxygen, and nitrogen – and an even smaller scattering of solid particles. Collectively, these solid particles are known to astronomers as "dust." But they are not at all like the stuff that collects on the floor under the bed. Interstellar dust particles are much smaller than everyday dust flecks and have a different make-up. Essentially they are grains of carbon "soot," with icy coatings of either

the orion nebula
This Hubble image of the very center of the Orion Nebula shows stars being formed within the brighter areas of the immense clouds of gas that make up the nebula.

frozen methane or water, or both. The ice freezes onto the grains because giant molecular clouds are very cold – only around 50–68°F (10–20°C) warmer than space itself. However, the distances between individual grains are truly staggering. Although these clouds contain enough material to make millions of stars like the Sun, they are so vast and thin that they are even emptier than the best vacuums scientists can create on Earth. They only look solid for the same reason that terrestrial clouds appear to be solid – we are looking at them from a very long way off.

remaining matter (less than 2%)

helium (25%)

molecular hydrogen (73%)

molecular composition
The percentage make-up of molecular clouds is represented here. They are composed mainly of molecular hydrogen, some helium, and tiny amounts of other matter.

gravitational contraction

Just as clouds on Earth change their shapes as they amble across the sky, so do molecular clouds as they drift through space. Earth clouds evolve with certain regions thinning out and becoming sparse while others become thicker and more opaque. Exactly the same thing is true of molecular clouds. Eventually, the molecular cloud develops several regions that have more material and are denser than their immediate surroundings. The gravitational influence of these dense cores is also stronger, which means that they attract nearby material and grow even tighter and more dense. This process, known as gravitational contraction, is the same process that built the first galaxies.

gravity rules
The law of gravity means that denser areas of cosmic material will attract more material and cause gravitational contraction to take place. This dark gas and dust cloud has a dense core that slowly pulls in nearby material.

dense core of gas and dust

gas and dust

As well as shrinking the original cloud, gravitational contraction has one other main effect. When gas is compressed, the size of the spaces between the individual atoms and molecules is also reduced, so collisions between them become more frequent. One way to visualize this is to imagine popcorn in a heated pan. If the pan were suddenly reduced in size, the corn kernels – having less space to move around – would hit the sides of the pan and each other more often. In a compressed gas, as with corn seeds in a small pan, the particles become more agitated, and the increased energy is detected as heat. In other words, as a molecular cloud core grows smaller and denser, it also gets hotter.

less space
Like popcorn in a small pan, the particles in a contracting molecular cloud collide and generate heat.

protostars

The first result of gravitational contraction is to produce a star cocoon – a small, dark blob of gas and dust less than a few light years across – which astronomers call a globule. Inside a globule, one or more stars are being formed. As the globule continues to contract it gets smaller and eventually gravitational energy is released in the form of heat and it begins to shine of its own accord. By this stage, the core has a surface temperature of several hundred degrees centigrade, and a central temperature a thousand times greater. Such a structure is not yet a star because it is still shrinking. It is known as a protostar.

protostar
The early stages in the formation of a star create a central region that is intensely hot, but which has not yet reached the temperatures of a true star in which fusion can occur.

stardom

The contracting protostar becomes even smaller and even hotter. The hotter an object becomes, the faster its atoms move. When the core of the protostar reaches a critical temperature of over 27 million °F (15 million °C), the hydrogen atoms in the core move so quickly that when they hit one another they stick together and create helium. Hydrogen fusion bombs work on the same principle, which is known as nucleosynthesis or nuclear fusion. Vast amounts of radiation are released, which marks a turning point in the object's formation. As with an exploding bomb, this new radiation exerts an immense outward pressure.

This force is so powerful that as the radiation powers away from the core, it holds in check the material further out that is trying to reach the center through the forces of gravity. This state of perfect balance between the outward pressure of radiation and the inward pull of gravity defines a true star. The nuclear reactor in the center making helium from hydrogen starts to shine steadily – and new light sets out on its journey from a new star.

protons (hydrogen nuclei)

helium nucleus (two protons and two neutrons)

energy created

starlight

the first stars

Stars come in different masses depending on how much material was available locally when the star was forming. The more massive stars have greater pressures on their centers due to the weight of material being attracted by gravity. This means that massive stars have to burn their nuclear fuel much faster

energy bomb
The fusion processes that provide immense quantities of energy in a star are the same as those at work in a hydrogen bomb.

nuclear fusion
Deep inside a star, four protons combine in nuclear reactions to make a helium nucleus. Two of the protons change into neutrons, so the final helium nucleus is composed of two of each particle. The process releases energy, the pressure of which keeps the star balanced against gravity.

British scientist **Sir Arthur Eddington** (1882–1944) was the first person to apply physics to the internal workings of stars. Thanks to the work of Albert Einstein, Eddington understood that vast quantities of energy were locked up inside even the smallest bits of matter. He also confirmed that light is deflected by massive celestial bodies, i.e. stars, just as predicted by Einstein's General Theory of Relativity.

to produce sufficient radiation to retain their structure, and they can run out of hydrogen in only a few million years. Smaller stars, such as the Sun, burn their hydrogen relatively slowly, and have lifetimes of around 10 billion to 13 billion years. Stars that are much less massive than the Sun can last many tens of billions of years, which is far longer than the current age of the Universe. It is likely that the Universe still contains stars that were among the first to form after the Big Bang.

the stars as our ancestors

However, most of the very first stars are long dead. The majority of these stars would have been extremely massive because of the vast quantities of star-making material that was freshly available from the relatively recent Big Bang. Being massive, they would have been short-lived, and would have had a very different make-up from the stars we see today. Although the stars that are forming now are still made almost entirely of hydrogen and helium, they also include some traces of heavier elements such as carbon, oxygen, and nitrogen, which are found in today's giant molecular clouds. This evidence shows that the first generations of stars must have been the factories of all the heavy elements found around us and within us.

❝ All of us are truly and literally a little bit of stardust. ❞

William Fowler, astrophysicist, 1983 Nobel Prize winner, (1911–1995)

As soon as the first stars formed, each one began converting hydrogen into helium in its core, just as stars do today. As the amount of helium within the center of a star steadily increases and the amount of hydrogen declines, the core gradually begins to cool and contract due to the lack of nuclear burning, and this leads to an increase in density. More frequent collisions among the nuclei in this confined space raise the temperature, and the fusion of helium nuclei is automatically ignited to create the highly unstable element beryllium. The density at the core of stars is so great that, despite existing for only a trillionth of a second, beryllium nuclei can collide with a helium nucleus and create a very stable nucleus of carbon. This helium-capture process can continue to construct heavier elements. For example, given a sufficiently high temperature, a carbon nucleus colliding violently with another helium nucleus will produce a stable oxygen nucleus. Each expansion and contraction cycle of a star increases the temperature and density at the core, allowing helium nuclei to be captured to form nuclei that are progressively heavier, such as magnesium and silicon.

globular cluster M22
M22 is an example of a globular cluster, which is a spherical island of stars found on the outskirts of galaxies, which can contain up to 10-billion years' worth of heavy elements that will be scattered through space when the stars die.

the first heavy elements

In this way, these first massive stars converted what was initially hydrogen and helium into a host of elements all the way up to iron in the periodic table. Iron, though, represents the endpoint of stellar nucleosynthesis. Whereas atoms lighter than iron give out energy when they are fused, keeping the star alive, iron atoms require energy to be fused – energy that a dying star lacks. And so, with no more fuel in the star to drive the reactions necessary to hold it up against gravity, any star with iron at its core is in big trouble. It succumbs to the inward pull of gravity, and it does so very suddenly, within seconds. This marks the end of the star's normal life.

star death

The star implodes so quickly that, within seconds, it reaches a very high density. The core recoils very briefly, which drives shockwaves out through the rest of the star at tremendous velocities and blasts much of the star's heavy elements into the depths of space. The energy released is tremendous and, for a while, the star shines brighter than an entire galaxy. This is the event that astronomers call a supernova, in which elements heavier than iron are forged. As powerful as it is, a supernova

star life and death

A region of gas and dust in a nebula contracts into a protostar, nuclear fusion begins, and a star appears. After millions of years, the supply of hydrogen runs out and how the star dies depends on its size. Small stars expand to form a red giant and their atmosphere is sloughed off as a planetary nebula. The remains become a white dwarf, which cools into an inert cinder known as a black dwarf. Massive stars expand into a red supergiant, then contract, and explode in a supernova, which leaves behind a neutron star, or in the case of immense stars, a black hole.

gas and dust in a nebula contract

protostar

star

does not always totally destroy the star. After the explosion, intense gravitational forces in the leftover core smash the electrons and protons together, and they cancel each other out. This leaves behind only the neutrons,

supernova blast wave
The Cygnus Loop Supernova is the remnant of a star explosion about 15,000 years ago. The red and yellow cloud is the shock wave from the explosion, and the blue ribbon of light could be gas blasted out by the supernova. It is thought to be traveling at about 3 million miles per hour (5 million kph).

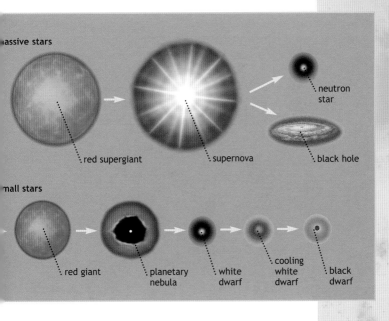

massive stars

red supergiant

supernova

neutron star

black hole

small stars

red giant

planetary nebula

white dwarf

cooling white dwarf

black dwarf

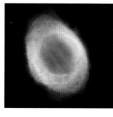

packed together like marbles in a bag, and the result is a city-sized sphere made entirely of neutrons: a superdense neutron star. But if the core is particularly massive, gravitational forces squeeze the object to breaking point. The star shrinks to a single dot, not even light is able to escape its gravity and it becomes a black hole. Not all stars end explosively though – only the more massive ones. The more lightweight stars, such as the Sun, lack the internal conditions that are required to create elements all the way up to iron. Instead, they expand to become vast and bloated stars with a pinkish tinge, known as red giants. Red giants evaporate into space, forming what is called a planetary nebula.

cosmic recycling

These cosmic detonations are the events that seeded the Universe with the first heavy elements. It is because of stars and supernovae that we have planets, mountains,

trees, and people – which are all made from the materials made in the stars and by supernovae.

Other debris left in space combines with the debris of other stars and assembles itself into molecular clouds – and the cycle of starbirth starts all over again. The Universe is the ultimate recycling machine.

building planets

Not so long ago, the only alien worlds definitely known to exist were the eight that, with the Earth, orbit the Sun. But in the past five or six years, that picture has changed completely. Thanks to recent advances in observing techniques, several dozen stars are now known to have planets going around them, just as our Sun does. Statistically speaking, it is now almost certain that planets are spread not just around the nearest stars, where most of the new planets have so far been found, but around a large proportion of all stars throughout space. Still, in the light of the latest astronomical research, this isn't really surprising because planets, it turns out, are natural by-products of the star-formation process.

proplyds

To recap, stars form when a vast cloud of gas and dust within a nebula starts to collapse in on itself under the influence of gravitational attraction. As the cloud shrinks, its center becomes more and more compressed, and gets steadily hotter until it shines of its own accord. However, as well as becoming more and more compact, the material destined to be a star also grows flatter by virtue of natural rotational motion. It is a bit like pizza dough flattening out as you spin it in the air. The result is that within about 100,000 years of the start of gravitational collapse, an enormous, swirling pancake of gas and dust, millions of miles wide, surrounds the newly forming star. Astronomers call such an object a protoplanetary disc, which is usually abbreviated to proplyd. Simply put, proplyds are planet factories. About 4,500 million years ago, the Solar System only existed as a giant disc – one that astronomers refer to as the Solar Nebula.

the planet mars
Like the Earth, Mars is a terrestrial planet – small and rocky with a diameter just over half that of Earth. Of all the planets, Mars' environment is the one most similar to our own. It is just possible that life might have existed there in the past – or even now.

planetesimals

At the very start of the planet-building process, a protoplanetary disc contains little more than gas molecules and grains of carbon dust with icy coatings orbiting the growing star in the centre. As the particles orbit the centre, they collide and are bonded together by electrostatic forces (the same ones that attract pieces of paper to a charged comb). In turn, these larger grains bond with others again and again and gradually their dimensions increase. Within a few thousand years, the dust grains grow to about the size of peas. After this, it is merely a question of a few hundred years before the disc is totally transformed. The result is a swirling tempest of countless asteroid-like bodies about half a mile (one kilometer) across. These "world-bricks" are known as planetesimals and they are the building blocks of planets.

The appearance of the first planetesimals represents a turning point in the planet-building process. These objects no longer have to rely on chance encounters with their orbiting neighbors to grow even larger. Instead, they have such substantial masses that they combine together by their mutual gravitational attraction. Growth through

building bricks
Asteroids (near right) are essentially planetesimals, the leftovers of the matter from which the planets were formed. In a sense they are planetary "building blocks," which could not coalesce to form a single planet because of the perturbing effects of the gravity of nearby Jupiter.

planet factory
The gas and dust cloud around Beta Pictoris is one of the best-known proplyds. The star in the center of the image has been blocked out, and false color has been applied to reveal the shape and extent of the surrounding cloud.

protoplanetary disc (proplyd) surrounding central star

dotted line shows central plane of disc

gravitational attraction is known as accretion, and finally turns planetesimals into true planets.

building planets

There is more to the story than this, though. The planets that form by the accretion of planetesimals are small and hard because they are formed from solid matter such as metals and silicates. These make terrestrial planets, such as Mercury, Venus, Mars – and Earth (which is the largest terrestrial planet in the Solar System). However, as we saw earlier, there are also much larger planets with very different compositions – these are the gas giants.

Gas giants start off in the same way as terrestrial planets, but they form relatively far from their parent stars. Their formation is a result of the temperature variation across a protoplanetary disc. Close to the center of the disc where it is denser, hotter, and closest to the star, only the heaviest materials – the rocks and metals – can condense from a gaseous state into the solid grains that seed the planet-building process. So the planets that form there tend to be terrestrial.

Neptune

little and large
Planets that are made of liquid gas, like Neptune, grow to be massive, while the terrestrial planets, like our own, are small and dense. Neptune is 38,800 miles (49,500 kilometers) in diameter, four times wider than Earth.

Earth

star (blocked out)

false-color indicates heat and density

planet formation

Planets form as a natural by-product of star birth. The initial step is the contraction under gravity of a giant molecular cloud. As it contracts, the rotation within the material flattens it out to form a disc. A star begins to form in the center, where it is hottest. In the rest of the disc, the material gradually accretes to form increasingly larger solid bodies. The ignition of the star expels the remaining gas and dust.

initial collapse of molecular cloud

protostar

material flattens into a rotating disc

gaseous material

gas and dust in disc accretes to form solid bodies

star ignites and the flood of radiation blows unused material away, leaving the star and planets.

However, at a certain distance from the center of the disc, a point is reached where the temperature drops low enough for gases, such as ammonia, carbon dioxide, and methane, to change their state and condense into solid ice. In interstellar clouds, these so-called "volatile materials" (which means that they change from a gas to a solid only at very low temperatures) are much more abundant than rock and metals, so the emergence and build-up of ice on the colder outskirts of a protoplanetary disc is considerable. All this extra volatile material on the fringes allows the icy planetesimals that form there to grow very large, and become perhaps 20 times more massive than a terrestrial world. Long before the rocky worlds closer to the center have finished growing, these larger worlds become so massive that their gravity pulls in hydrogen and helium gas directly from the disc as well as pulling in other orbiting planetesimals. Terrestrial worlds never become massive enough to exert this gravitational influence. What start off as balls of rock and ice far from the central star eventually become vast spheres of compressed gas enshrouding solid cores – gas giants.

The German-born English astronomer **Sir William Herschel** (1738–1822) found the first planet in the Solar System orbiting beyond Saturn. He named this planet Uranus. With the aid of a 4-foot (1.2-meter) telescope that he built – then the largest telescope in the world – Herschel went on to find the planet's largest two moons, and also two of Saturn's.

the evidence

Our Solar System is a good starting point to look for evidence to support this theory of planet formation. As expected, the worlds closest to the Sun are all small and made of rock and metal. Also the giant planets – Jupiter, Saturn, Uranus, and Neptune – are found further from the Sun. And there is other evidence too. Between the orbits of Mars and Jupiter lies a swarm of tumbling debris – the asteroids. These are essentially planetesimals left over from the planet-building process. Similarly, beyond Neptune there is another band of accumulations of ice debris, known as the Kuiper Belt. Pluto, the Solar System's oddball planet, is often considered to be a large Kuiper Belt object rather than a real planet. Lastly, the Solar System is essentially flat, with all the planets (except Pluto) orbiting in the same plane around the Sun, and in

asteroid belt
The asteroid belt is a ring of irregular objects. Most asteroids orbit in this belt between the orbits of Mars and Jupiter. Over 10,000 asteroids have so far been discovered and are of varying sizes.

sun

mars

asteroid belt

the same direction. This is exactly what would be expected if the Solar System had formed from a disc.

Further afield, astronomers have dug up even more convincing evidence that their theories of planet formation are correct: they have actually seen protoplanetary discs for themselves. One of the most famous examples is the one surrounding a star called Beta Pictoris in the constellation of Pictor (see pages 44–5). The disc is seen almost perfectly edge on. More recently, the Hubble Space Telescope has found several proplyds in the giant Orion Nebula, in the constellation of the same name.

> **"A time will come when men will stretch out their eyes. They should see planets like our Earth."**
>
> Christopher Wren (1657)

detecting other planets

As well as locating and identifying protoplanetary discs, in the last few years astronomers have also found dozens of planets that are orbiting other stars. These discoveries have finally closed the lid on the question of whether planets are common or rare. However, this astronomical detective work has been far from easy, which is why astronomers have taken so long to prove the existence of real alien worlds. The problem is that no equipment can actually see these planets directly – at least, not yet.

center of mass
As stars are much more massive than planets, the center of mass, around which both objects revolve, is usually inside the star itself (see top illustration). But the more massive the planet (bottom illustration), the closer the center of mass is to the planet.

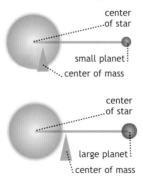

center of star

small planet

center of mass

center of star

large planet

center of mass

Planets do not shine of their own accord. They reflect the light of their parent stars as a dim glow that is easily rendered invisible by their parent star. So as these extrasolar worlds, (as they are called), generally cannot be seen, astronomers must detect them by their effects on their parent stars.

Strictly speaking, planets do not orbit stars. Instead, the star and planet orbit their mutual center of mass – the point where they would balance if you joined them with a stick. And because stars are very much more massive than planets, the center of mass is usually within the star itself. So while the planet follows a big circle around the center of mass, the star merely wobbles back and forth like a child with a hulahoop. This is the motion that reveals the presence of the planet. If the star had no planet going around it, it would simply rotate around a perfectly still central axis that didn't wobble.

possible gas giant

filament

binary protostars

extrasolar giants

Thanks to spectroscopy (see page 15), astronomers can see how stars move, and observation of these wobbles led to the first discoveries of extrasolar planets, in 1995. To date, several dozen wobbly stars are known to be orbited by planets. In most cases only one planet is indicated. But some stars definitely have more than one planet, and perhaps they all do (smaller planets could also be present, but as they create only small wobbles, they may have simply escaped detection). Currently, spectroscopy can only detect the motion caused by very large planets. All the extrasolar planets found so far are giants, usually even more massive than the Solar System's largest planet, Jupiter. However, techniques are improving and the equipment is becoming more sensitive all the time. Already, planets that are intermediate in mass between Jupiter and Saturn have been indicated and it is only a

gas giant
This Hubble Telescope image is possibly of a planet known as a gas giant. Initial ideas are that two protostars have begun to orbit one another as binary stars, and the gravitational effect has been to eject the gas giant from its orbit around one of the two stars. The filament of light could be the trail of the planet as it begins wandering in space.

question of time before astronomers find even smaller worlds. However, it is doubtful if the method of detecting planets by stellar wobbles will ever be sensitive enough to find anything except large planets; if they are to find other earths, astronomers will need other techniques.

other earths

One method that looks likely to yield results is to search

for planets in so-called transit. During a solar eclipse, the Moon passes in front of the Sun and as a result it gets dark on Earth. Similarly, the planets Mercury and Venus periodically pass in front of the Sun – passages known as transits. Now picture an extrasolar planet orbiting a distant star, and imagine that we see the orbit almost edge-on. Eventually, the planet passes in front of its star, causing the light levels to dip slightly and return to normal once the transit is over. Some time later, when the planet has completed another orbit of its star, it makes another transit. And then another, later still. This produces a periodic series of dips in the star's light intensity. Although these changes are minute, astronomers can detect them with the latest equipment. Again, this method is more sensitive to the more massive and larger planets, but it could conceivably reveal much smaller worlds.

planet in transit
Seen from the Earth, a planet passes in front of its parent star, resulting in a tiny reduction in the star's normal light intensity. Astronomers can detect these changes and infer the presence of planets, even if they cannot see them.

Another method is to search for the light that planets reflect from their stars. However, planetary light levels are beyond detection by any instrument available today – except for those from large planets far from their stars and not lost in their glare. But by linking large telescopes on land or in space, it is possible to improve dramatically the overall resolution. Plans are now being made to construct planet-finding arrays of telescopes in space.

birth of life

Life on Earth is everywhere. We find it on land, in the air, in oceans at depths where sunlight cannot penetrate, and even in solid rock. In short, life is found under an amazingly diverse range of conditions. The first primitive life forms arose as soon as the planet cooled down. Since then, these primitive cells have evolved to the complexity we see today. Despite this rich variety, organic life still occupies only one single world. There is no convincing evidence for life elsewhere, either in our Solar System or on planets around distant stars. Some astronomers insist that life – even intelligent life – is common throughout the Universe. But we simply do not know for sure.

defining life

Anyone can tell the difference between living things and inanimate objects. But when it comes to actually defining what it means to be alive, even the greatest scientists find it hard to supply an answer. Still, there are a few traits that only living things possess, and a range of functions that only living systems perform. All living organisms, whether plants or animals, take in energy from their surroundings and use it to nourish themselves and to grow. They can reproduce themselves, they can respond to external stimuli, and they can adapt to meet the conditions necessary to survive in a changing

carbon-based
All known life forms are carbon-based, but that does not mean that all carbon-based forms are necessarily alive. The ordinary pencil contains carbon, but only in inorganic form.

living animal   living plant   inanimate objects

environment. Lastly, all life forms that we know of are composed of molecules based on the element carbon. This is because carbon, more than any other element, has the ability to form very massive, yet stable, molecules including deoxyribonucleic acid

The first person to suggest that life on Earth may have been seeded from space was the Swedish chemist **Svante Arrhenius** (1859–1927). Arrhenius believed that life came from tiny organisms embedded in the dust particles found in comets, brought to Earth when those comets crashed into our planet long ago.

signs of life?
This section of rock from Mars has been greatly magnified and captured in a false color image. Some scientists think that the blue form may be fossilized bacteria, indicating primitive Martian life.

(DNA). The DNA molecule can be thought of as a chain of information and instructions contained in every living cell in an organism. These instructions tell the cell how to use energy from its surroundings and to perform its individual role within the organism as a whole.

seeding

The oldest known microfossils date from around 3.5 billion years ago, less than a few hundred million years after the Earth had formed. It would seem then

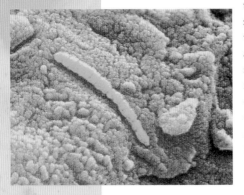

that primitive life appeared as soon as the planet had cooled to habitable levels. This rapid appearance has led some scientists to believe that the necessary ingredients – the so-called organic molecules made of hydrogen, carbon, oxygen, and nitrogen – were present from the beginning and didn't form on Earth. If so, from where did these organic molecules come? One possible answer is that the Earth was seeded from outer space. Astronomers have already found dozens of organic molecules embedded in the interstellar dust clouds from which stars and planets are formed. Examples include hydrogen cyanide and formaldehyde. It could be that the nebula from which the Solar System sprang was full of these ingredients, and that these seeded the Earth and paved the way for biochemical reactions.

Alternatively, some astronomers have suggested that these complex molecules fell to Earth on comets and meteorites during its formation. And still others believe that these comets and asteroids contained not merely organic molecules, but actual life forms, such as simple bacterial spores, formed on distant planets and carried to Earth when chunks of these planets were blasted off by meteorite impacts. However, it is more likely that the very early Earth harbored only organic molecules rather than actual life forms from space.

possible soup ingredients
Dust clouds are known to contain simple organic compounds, which could reach Earth. Amino acids could have evolved from carbon-based molecules created in the upper atmosphere by radiation, which then fell into the early oceans. Bacteria are known to form spores that can lie dormant for millions of years, possibly while they travel through space.

interstellar dust cloud . amino acids . bacteria .

first life forms

Irrespective of how these organic molecules arose, they swiftly organized themselves into more complex forms. In the oceans, the organic molecules were diluted with water to form what scientists often refer to as a primordial soup. Driven by the heat of the young planet's formation, some of these molecules, called amino acids, joined together to make chains of molecules called proteins. Eventually a molecule developed that could make simple copies of itself – it could reproduce. Then, in order to ensure its survival, a reproducing molecule arose that surrounded itself with a wall of protein to form the first-ever protective cell membrane. The ability to reproduce and the instinct to protect and survive are definite indicators of life. In a very short space of time, geologically speaking, the primordial soup gave birth to the first single-celled life forms on Earth.

evolution

These first organisms would have looked like algae or bacteria. All life that exists on the planet today came from these first simple spores. The key to this great metamorphosis and diversification is evolution, which is the process of life adapting to its changing environment and, in doing so, becoming "fitter" – better equipped to survive – and usually more complex. Evolution is driven by gene mutation and natural selection. To see how it works, imagine a herd of giraffes living off the trees in a forest. No offspring is an exact replica of its parent – because of slight differences or random mutations in their genes. That is why some mature giraffes have shorter necks and others have longer necks. If the trees in the region were to grow taller, low-level branches would become more scarce. Those giraffes with shorter necks would not be able to reach the higher branches. They would face a food shortage and many might starve to death. Over the generations, with many of the short-necked giraffes dying before they reproduce, their particular gene mutation would become increasingly rare. The better-adapted long-necked giraffes would become the ones that survive and would be able to

a tall story
Darwin's theory of evolution showed that animals do adapt to their environment and that changes occur as a result of random mutation, modified by the process of natural selection.

reproduce and pass on their long-neck genes.
With short-necked giraffes "selected against," as
scientists say, the population would gradually
evolve longer necks.

life elsewhere?

Life is now infinitely more
complex than when it first
appeared. But what about
life elsewhere? Does it exist?
The answer is that we simply
don't know. Some scientists
point out that life on Earth
took hold as soon as it was
able to, and that the same
events must have occurred
elsewhere. But others
claim that the first self-

meeting aliens
*There is no real
evidence that
aliens exist, but
they are often
depicted in
fiction in
myriad
strange
forms.*

replicating molecules arose by the tiniest of flukes, and it
is unlikely that the same thing will happen elsewhere.

Nevertheless, if we do assume that life is spread across
the entire Universe, what are the chances that any of
these alien life forms are intelligent? The answer to that
depends on evolution. Evolution doesn't only change
physical traits such as the length of an animal's neck or
the speed with which it can flee a predator. Intelligence
also has to evolve. Any early hominid with the wits to
pick up a branch to use as a weapon against a saber-
toothed tiger would have a survival advantage over the
hominid who simply tried to outrun a predator. So it
can be argued that if life has arisen elsewhere and has
survived, then it would have necessarily developed a form
of intelligence. In recent years, astronomers have been
using radio telescopes to listen for messages from space.
Some are optimistic that, if these civilizations do exist
somewhere, we will find them within a few decades.

key points

• All life forms
are composed of
molecules based
on carbon.
• The first life
forms on Earth
developed from a
"primordial soup."
• Evolution is the
process of life
adapting to
its changing
environment.

into the
unknown

Astronomy is a dynamic and changing science in which new discoveries can lead astronomers into the unknown. Technological advancements yield more precise observations, as well as those that were once technically impossible. These new advances often mean that old theories are discarded and new ones embraced – until they, too, need to be abandoned or revised. Here are some of the latest ideas about the Universe we live in. The relatively new idea that not only is the Universe expanding, but that it is also expanding at a faster rate, is introduced. And if we are living in an accelerating Universe, then an explanation is needed for this acceleration – at the moment, the most likely cause is believed to be dark energy. An alternative to the accelerating Universe is the idea that the speed of light is changing over time. As well as looking at these ideas, we also look at baby universes, wormholes, time travel, and the final question of how the Universe might end. Although many of these ideas may seem bizarre, they are taken very seriously in cosmology.

the search for life in space
The Search for Extraterrestrial Intelligence – SETI – is a project carried out by volunteers everywhere who use their computers to examine these records of radio emissions for signal patterns that would imply a source that is intelligent.

nature of the universe

Every once in a while, a new theory or piece of data comes along that completely changes how astronomers view the Universe. Once, it was thought that space consisted only of planets, stars, and galaxies. Then came the revelation that much, indeed most, of the Universe's material is undetectable – for every piece of mass we can see, there is about 10 times more dark matter that we cannot. This discovery forced a radical rethink of old assumptions in many areas of astronomy. Now, another discovery is having perhaps an even greater effect on scientists' thinking. Could the rate at which the Universe is expanding actually be accelerating?

accelerating universe

Astronomers know that the Universe is expanding. This means that the background radiation that fills the Universe has been stretched or "redshifted." And the light from distant galaxies has experienced the same stretching effect. But the latest results are hinting that the expansion is actually speeding up. Normally, the further away a

getting larger faster
A Universe that is expanding at an increasing rate doubles in size in shorter and shorter periods of time.

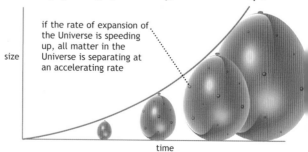

size

if the rate of expansion of the Universe is speeding up, all matter in the Universe is separating at an accelerating rate

time

spacetime

In this theory, space and time are bound together to make a single four-dimensional entity known as spacetime. Now, instead of representing the position of an

a model for gravity
Spacetime is like a rubber sheet, with stars as heavy balls. The ball warps the rubber and deflects the paths of smaller objects so that they move, by gravity, towards it.

object by three spatial coordinates (height, breadth, and depth), its position is represented with the addition of a fourth coordinate – one of time. So the location of an object in spacetime depends not only where it is, but also when it is there. Astronomers say that this spacetime is curved, and it is more curved in the vicinity of large, compact masses such as black holes.

heavy ball bends the sheet to create gravity

curved sheet represents spacetime

galaxy is, the more time it takes its light to reach us. The Universe continues to expand after the light leaves the galaxy, and so the more redshifted the light will be when it reaches Earth. Measure the redshift and you measure the distance. But this assumes that the Universe is expanding at a constant rate, and this may not be true. Let's suppose it had been expanding more slowly in the past. If this were so, the light from distant galaxies would have been redshifted by smaller amounts. Remember that a smaller measured redshift implies a smaller distance. In other words, if you assume that the expansion was slower in the past, then the distances to faraway galaxies will all be underestimates. Support for the idea of an accelerating Universe has been obtained from observations of supernovae in very distant galaxies. They all look much too dim to be as close as their redshifts suggest.

dark energy

Soon, more atypically dim supernovae were detected, which added weight to the controversial idea of an increasingly faster-moving Universe. What was needed next was an explanation for what was causing the expansion to speed up. For that, some scientists came

up with a concept called dark energy. They claim that the Universe is filled with a mysterious energy field that acts like a large-scale repulsive force or negative pressure, which is strong enough to counteract gravity. While gravity tries to pull all the matter in the Universe together, the dark energy not only resists this pull, but causes spacetime to spring apart, faster and faster. If this is so, then up to 70 percent of the Universe's mass could be embedded, in the form of energy, in the very fabric of spacetime. That is, spacetime itself might have mass. The matter that we can actually see, in the form of stars, galaxies, nebulae, and planets, might only amount to a few percent of the total matter in the Universe.

variable light speed

An alternative theory to an accelerating Universe is the idea that the speed of light is slowing down. Here's how it works. Suppose light traveled 10 percent faster in the past than it does now. If so, then the light from a galaxy 10 billion light years away would reach us in 9 billion years. Because space is expanding, redshift increases with time,

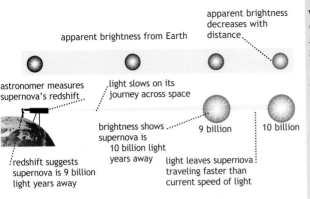

apparent brightness from Earth

apparent brightness decreases with distance.

astronomer measures supernova's redshift.

light slows on its journey across space

redshift suggests supernova is 9 billion light years away

brightness shows supernova is 10 billion light years away

9 billion

light leaves supernova traveling faster than current speed of light

10 billion

and the light from the galaxy will be redshifted by a figure corresponding to the expansion of the Universe over that 9 billion-year journey. Astronomers, measuring the redshift, will infer a distance of 9 billion light years. However, because the distance to the galaxy is actually 10 billion light years, the galaxy will appear dimmer than it should be according to its inferred distance. Thus, some cosmologists have suggested that the recent anomalous supernovae (pictured opposite) might appear dimmer than their measured redshifts suggest not because the space is expanding more quickly now, but because the speed of light is slower now than it was. Both explanations produce the same result.

In 1917, **Albert Einstein** (1879–1955) concluded from his General Theory of Relativity that the Universe is expanding. This appeared ludicrous at the time, so he made an amendment, called the cosmological constant, to his equations to make the Universe static. Later, when Edwin Hubble confirmed observationally that space was expanding, Einstein regretted his earlier position, calling it his "greatest blunder."

baby universes

Another recent idea is that our Universe may be one of an infinite number of

baby universes
Physicists are exploring the idea that our own Universe may be connected to smaller, baby universes by wormholes or spacetime tunnels.

········ wormhole or tunnel

baby universe ············

universes. According to Einstein's General Theory of Relativity, a black hole could be a gateway to another universe and any matter swallowed by the hole could pass through and enter an "external" region of spacetime. As this new spacetime "bubble" buds off from our own Universe, cosmologists refer to it as a baby universe. Every time a black hole appears in our Universe, it could lead to the birth of another baby universe, which then grows to maturity. Our Universe might be connected, via black holes, to innumerable other universes – and each one might be connected to others. It is possible that our Universe was once a baby universe, created when a black hole formed in another universe. Many scientists take the idea seriously – it is one consequence of the mathematics of general relativity.

wormholes and time travel

A consequence of the same mathematics are wormholes – tunnels that lead from a black hole in one universe to a connecting universe or to a different place in the same universe. So the material that falls into a black hole could exit in the same universe, but in a different place. Not only that, but the time period at one end of the wormhole is not necessarily the same as that at the far end. The idea raises the possibility of traveling rapidly to distant parts of the universe, as well as backward and forward in time, although this is highly unlikely.

one-way trip
One of the drawbacks of using a wormhole for space and time travel is that anything with mass would cause the wormhole to close extremely quickly, which may rule out a return journey.

end of the universe

We opened this book with the question: where did the Universe come from? Now it is time to look to the other extreme, and contemplate the Universe's ultimate fate. There are three possible endings to the Universe – or at least, so it currently seems. Either the Universe will go on expanding forever, or it will one day reverse and start to collapse, or it will oscillate between expansion and contraction forever. Which of these actually takes place depends on only one thing: the total gravity of all the matter in the Universe.

critical density

Even though the Universe is expanding, its gravity is constantly trying to pull everything back together again. A rocket launched on Earth has to travel at nearly 7 miles per second (11 kilometers per second) to escape from the planet and head into space. If it fails to reach this velocity, the rocket falls back to Earth. But a rocket launched from Jupiter has to reach much faster speeds if it is to escape because Jupiter has a higher gravity. Now try to picture the expanding Universe as a rocket launched from a planet. Exactly as the planet's gravity acts on the rocket and tries to reverse its motion, so the gravity of all the matter in the Universe itself acts to try to make all the galaxies reverse their direction. Just as with the rocket, whether gravity wins the day depends on the ratio between the strength of the gravity and the amount of upward thrust. In addition, whether the gravity of the Universe is sufficient to reverse the

goodbye gravity
To escape from Earth's gravity, the Saturn V rocket, which took astronauts to the Moon, had five engines that used 15 tons of fuel a second to generate 3,400 tons of thrust.

expansion depends on the density of the Universe – how close together the galaxies are. If the density of the Universe is lower than a critical density, gravity is too weak and the Universe expands forever.

open or flat universe?

Cosmologists talk about an "open" Universe when they are referring to one that expands forever. In this type of universe there is not enough matter to bring the expansion to a halt. A "flat" Universe is also expanding forever, but is always just on the verge of stopping. Both sound everlasting, and in some ways this is true. But although the Universe may last forever, the planets, stars, and galaxies within it will not. As more and more stars are born and die, the raw materials in the Universe slowly run out. One day, the last star will die – and there will be no more matter to make others. The dead stars that remain will simply cool down and fade from view – not that there will be anybody around to view them. This is grim, but the ultimate fate is even worse, because not only are the stars and galaxies living on borrowed time, so too are the subatomic particles from which they are made – all are inherently unstable in the long term. At some

open and flat universes
An open universe expands forever and heads towards complete emptiness. If the flat model of the Universe is correct, then it must have expanded extremely rapidly during a period early in its history.

open universe

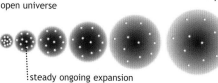

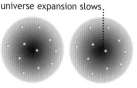

universe expands forever.

steady ongoing expansion

flat universe

universe expansion slows.

rapid early expansion

time in the far distant future, if we live in a flat or open Universe, all matter will decay into radiation, and cease to exist. Spacetime itself will go on, but there will be nothing left in it.

the big crunch

Given the recent thinking that the Universe's expansion is speeding up, it is very possible that we do live in an open Universe. The alternative possibility is that the Universe's

closed universe

universe expands

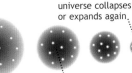
universe collapses or expands again.

universe contracts

density is greater than critical and the Universe is "closed." Here, gravity will one day win and everything then starts to collapse inward. In this case, two ends are possible. The Universe could collapse back to a single point – mimicking the Big Bang in reverse – and then cease to exist. This situation is often referred to as the big crunch, where the endpoint of infinite density, called the Omega Point, is reached. Alternatively, the Omega Point might never be reached. In this model, the contraction one day reverses direction again, and the Universe resumes its expansion, cycling endlessly between contracting and expanding states.

Whichever of these possible situations is applicable to the real Universe, time is definitely on our side. As far as we can tell, armed with our current understanding of the universe, nothing drastic is going to happen to our cosmos for a very, very long time indeed.

closed universe
A closed universe expands, then contracts. This may lead to the big crunch, or the contraction may reverse and the universe will start expanding again.

is this where we came in?
A closed universe is destined to collapse in on itself; the only uncertainty concerns how close it will come to returning to the point from where it all began.

glossary

absolute zero
The lowest physically attainable temperature. It corresponds to –459.69°F (–273.16°C).

accretion
The process whereby planets are built from progressively larger fragments in a disc around a young star.

antimatter
Material whose fundamental particles have properties, such as electric charge and spin, which are reversed compared with normal matter. For example, an electron has an antiparticle equivalent, known as a positron.

asteroid belt
The band of rocky and iron asteroids found between the orbits of Mars and Jupiter.

atom
The smallest unit to which an element can be reduced and still be recognized as that element. Atoms are composed of a positive core orbited by negative electrons.

baby universe
A universe that buds off from another universe through a spacetime tunnel, and is perhaps entered through a black hole.

big bang
The event in which space, time, and matter were all created 12 to 15 billion years ago.

black hole
A region of space with such powerful gravity that nothing, not even light, can escape from it.

blueshift
The shift in wavelength of a light source toward the blue region of the spectrum that is apparent when the emitting object is approaching the observer. It is caused by the Doppler effect.

carbon
The fourth most abundant element in the Universe. Carbon is important in nucleosynthesis, and forms the basis of all known life.

cepheid variables
A star that periodically brightens and dims, and by comparing that period with the star's brightness, its distance can be established.

cluster
A group of stars or galaxies held together by gravity.

comet
A small and irregular body made of dirty water ice. As a comet approaches the Sun, its ices melt and trail off into space to form a long cometary tail.

cosmic background radiation
The omnipresent microwave radiation that represents the remnant of the radiation that filled the early Universe.

dark energy
A form of energy which is believed be embedded within space itself and is causing the expansion of the Universe to accelerate.

dark matter
An indirectly detectable form of matter, but which astronomers know must exist because of its gravitational influences.

DNA
The molecule of life, made chiefly from carbon, which instructs individual cells how to use energy to perform their specific role in an organism.

doppler effect
The shift in wavelength of sound or light caused by relative motion between the observer and the source.

electromagnetic spectrum
The term used to describe the entire band of radiation that includes radio waves,

infrared, light, ultraviolet, X-rays, and gamma rays.

electron
A negatively charged subatomic particle, usually found orbiting the nucleus of an atom.

galaxy
A vast collection of stars held together by gravity. Galaxies are either spiral, elliptical, or irregular.

gas giant
A planet made chiefly of hydrogen and helium. Gas giants include Jupiter and Saturn, and most of the planets so far found around other stars.

globular cluster
A roughly spherical cluster of very old stars held together by gravity.

globule
A dark nebula in which star formation is occurring.

gravity
The force that holds everything in the Universe together. In relativity, gravity manifests itself as a curvature, caused by the presence of mass, in the fabric of spacetime.

helium
The second most abundant element in the Universe and one of three created in the early Universe.

hydrogen
The most abundant element in the Universe, created in the Big Bang. Hydrogen is

found inside stars and nebulae. About 73 percent of all visible matter is made of hydrogen.

inflation
The period during which the very early universe went through a process where it grew rapidly to roughly the diameter of a melon.

ionization
The process whereby energetic photons knock electrons away from atoms and render the atoms positively charged. The resultant atoms and the free electrons are then known as ions.

kuiper belt
The band of cometlike bodies orbiting the Sun beyond Neptune.

lithium
The third most abundant element in the Universe, created in small quantities along with hydrogen and helium in the Big Bang.

local group
The cluster of galaxies to which our galaxy, the Milky Way, belongs.

luminosity
The amount of radiation a star, galaxy, or other astronomical object emits per second.

molecular cloud
A dark nebula made chiefly of hydrogen gas in which the individual atoms are paired to make molecular hydrogen.

molecule
A group of atoms held together by sharing their electrons. Carbon dioxide, for example, at the atomic level is a molecule consisting of one atom of carbon and two of oxygen.

nebula
A cloud of gas and dust in space, from which stars form. Nebulae are also created when stars die.

nucleosynthesis
The process whereby elements heavier than lithium are created inside stars as a result of nuclear reactions.

organic molecule
A molecule with a basic skeleton made up of atoms of carbon, oxygen, and hydrogen.

photometry
The science of measuring the amount of light an object emits, and observing changes in its light output over time.

photon
A particle of radiation from the electromagnetic spectrum. Light rays, for example, can be considered a stream of photons.

planet
Any substantially sized object that orbits a star, but which is not an asteroid or a comet.

planetary nebula
A nebula, often symmetrical, created when

a sunlike star throws off its atmosphere at death.

planetesimal
An asteroid-like body formed by accretion in a disc around a young star. Further accretion turns planetesimals into planets.

proplyd or protoplanetary disc
A disc surrounding a young star in which planets will one day form, or are already forming, via accretion, and are orbiting the central star.

protostar
A large, cool object that is contracting under gravity, which when it becomes sufficiently compact will initiate nuclear reactions and shine as a new star.

quark
The most fundamental form of matter that physicists currently know of. Quarks are far smaller than atoms, and it is thought that they are completely indivisible.

radiation
The emission of electromagnetic waves, (such as light waves, radio waves, heat, X-rays, or gamma rays), or a stream of particles, (such as electrons, neutrons, protons, or alpha particles), from an energy source, and the transmission of these waves or particles through space or through a material.

red dwarf
A very common type of low-mass star, roughly half the size of the Sun and considerably dimmer.

red giant
A giant star many hundreds of times larger than most stars. The Sun will become a red giant when it dies.

redshift
The shift in wavelength of a light source toward the red region of the spectrum caused as the emitting object moves away from an observer. It is caused by the Doppler effect.

singularity
A region in spacetime that has no physical size, but possesses infinite density. The Big Bang formed from a singularity.

spacetime
According to relativity, space and time are not separate entities, but are two parts of the same thing: a spacetime continuum.

spectral lines
Dark or bright vertical lines in a spectrum caused by the absorption or emission of light at specific wavelengths.

spectroscopy
The study of the spectrum of light from astronomical and other bodies.

spectrum
The rainbowlike band obtained when an object's light is passed through a prism or a fine grill.

star
Any object, including the Sun, which shines by, and obtains energy from, nuclear reactions at its core.

supercluster
An agglomerations of clusters of galaxies. They are the largest gravitationally bound objects in the known Universe.

supernova
An exceedingly bright event that marks the explosion of a giant star at the end of its life.

terrestrial planet
A planet, such as Earth, made from rocky and metallic substances, and not from gases.

white dwarf
The dead, compact core of a sunlike star, left behind after the star sheds its atmosphere in a planetary nebula.

WIMP
A Weakly Interacting Massive Particle. WIMPs are fundamental particles that may make up some of the Universe's dark matter.

wormhole
A bridge between two places in the same universe or in two different universes. Travel though a wormhole, if feasible, opens up the possibility of time travel or faster-than-light travel.

index

Further reading

The Story of the Solar System
Mark A. Garlick, Cambridge
University Press, 2002

Foundations of Astronomy
Michael A. Seeds,
Wadsworth Publishing
Company, 2000

Companion to the Cosmos
John Gribbin, Phoenix
Giant, 1996

*Big Bang: The Story of the
Universe* Heather Couper
and Nigel Henbest, Dorling
Kindersley, 1997

Related websites

http://seds.lpl.arizona.edu/n
ineplanets/nineplanets/nine
planets.html *A multimedia
tour of the Solar System*

http://astronomylinks.com/
Astronomy and space links

http://www.kidsastronomy.
com/*A kids' guide to
astronomy*

http://www.astronomytoday
.com/Articles and links on
astronomy, cosmology, and
space exploration

http://www.space-art.co.uk/
*A collection of artworks
covering all aspects of
astronomy, each with complete
scientific captions*

Acknowledgments

I would like to thank John Gribbin for approaching me with the idea for writing this book. I would also like to thank Dorling Kindersley for commissioning me to do just that, and all at Design Revolution for putting the book together. Lastly, thanks to Delphine for her encouragement, support, and understanding. She's a star brighter than any in this book.

Illustration
Richard Tibbitts and Martin Woodward, AntBits illustration

Additional Illustration
Joerg Colberg 11

Additional picture research
Penni Bickle

Index
Indexing Specialists, Hove

Picture credits

Bettmann/Corbis: 21, 32(b), 47(t), 61(b). **Corbis Images:** Philip Gould 16, 17(tr); Roger Ressmeyer 26. **Hulton-Deutsch/Corbis:** 20(b), 38, 52(t). **Mary Evans Picture Library:** 55. **NASA and The Hubble Heritage Team:** 4; 9 (br); 10; 13(bl); 13(br); 18; 19(c); 27(t); 28(b); 33(b); 34; 39(t); 42(t); 44(tl); 44(b); 45(c); 45(b); 53; 56; Matt Bobrowsky 5(c); Wolfgang Brandner, Eva K. Grebel, You-Hua Chu 8; R. Evans, and K. Stapelfeldt, 45(t); Don Figer 32(t); A. Fruchter 42(b); J.J Hester 41(t); J.C Howk and B.D Savage 28(t); Adam Riess 60; Martino Romaniello 7; K.Sahu, S.Casertano, M.Livio, R.Gilliland, N.Panagia, M.Albrow, and M. Potter 39(b); S. Terebey 49. **Science Museum:** 13(tl). **Science Photo Library:** 30; Lynette Cook 47(b); Michael Dunning 20(t); Nasa 52(b); US Navy/Science Photo Library 37(t). **SOHO–EIT Consortium:** 14a, 14b, 14c.

Every effort has been made to trace the copyright holders.
The publisher apologizes for any unintentional omissions and would be pleased, in such cases, to place an acknowledgment in future editions of this book.

All other images © Dorling Kindersley.
For further information see: **www.dkimages.com**